50 THINGS TO KNOW BOOK SERIES REVIEWS FROM READERS

I recently downloaded a couple of books from this series to read over the weekend thinking I would read just one or two. However, I so loved the books that I read all the six books I had downloaded in one go and ended up downloading a few more today. Written by different authors, the books offer practical advice on how you can perform or achieve certain goals in life, which in this case is how to have a better life.

The information is simple to digest and learn from, and is incredibly useful. There are also resources listed at the end of the book that you can use to get more information.

50 Things To Know To Have A Better Life: Self-Improvement Made Easy! by Dannii Cohen

---

This book is very helpful and provides simple tips on how to improve your everyday life. I found it to be useful in improving my overall attitude.

50 Things to Know For Your Mindfulness & Meditation Journey by Nina Edmondso

---

Quick read with 50 short and easy tips for what to think about before starting to homeschool.

50 Things to Know About Getting Started with Homeschool by Amanda Walton

50 Things to Know

I really enjoyed the voice of the narrator, she speaks in a soothing tone. The book is a really great reminder of things we might have known we could do during stressful times, but forgot over the years.

- HarmonyHawaii

50 Things to Know to Manage Your Stress: Relieve The Pressure and Return The Joy To Your Life

by Diane Whitbeck

There is so much waste in our society today. Everyone should be forced to read this book. I know I am passing it on to my family.

50 Things to Know to Downsize Your Life: How To Downsize, Organize, And Get Back to Basics

by Lisa Rusczyk Ed. D.

Great book to get you motivated and understand why you may be losing motivation. Great for that person who wants to start getting healthy, or just for you when you need motivation while having an established workout routine.

50 Things To Know To Stick With A Workout: Motivational Tips To Start The New You Today

by Sarah Hughes

50 THINGS TO KNOW ABOUT ARCHAEOLOGY

A BRIEF INTRODUCTION

JL Musgrave

50 Things to Know

50 Things to Know About Archaeology   Copyright © 2018 by CZYK Publishing LLC. All Rights Reserved.

All rights reserved. No part of this book may be reproduced in any form or by any electronic or mechanical means including information storage and retrieval systems, without permission in writing from the author. The only exception is by a reviewer, who may quote short excerpts in a review.

Cover designed by: Ivana Stamenkovic
Cover Image: https://pixabay.com/en/foro-romano-rome-antiquity-italy-3676169/

Edited by:

CZYK Publishing Since 2011.

50 Things to Know
Visit our website at www.50thingstoknow..com

Lock Haven, PA
All rights reserved.

ISBN: 9781726725835

50 THINGS TO KNOW ABOUT ARCHAEOLOGY

50 Things to Know

BOOK DESCRIPTION

Are you a fan of Indiana Jones?

Have you always been curious about what archaeologists really do?

Want to learn more about the world's most significant archaeological wonders?

If you answered yes to any of these questions, then this book is for you...

50 Things to Know About Archaeology by JL Musgrave offers an approach to archaeology that is simple and concise. Most books on archaeology tell you information in bulk. Although there's nothing wrong with that, 50 Things to Know About Archaeology answers your questions about archaeology in simple, concise, short paragraphs. Based on knowledge from the world's leading experts, JL Musgrave utilizes academic and public resources to simplify archaeological information for the novice enthusiast.

In these pages you'll discover what archaeology really is, the tools of the trade, and the significant archaeological sites that shaped the way we interpret the past. This book will help you establish a foundation of knowledge to build upon as desired.

By the time you finish this book, you will know what archaeology truly is. So, grab YOUR copy today. You'll be glad you did.

50 Things to Know

TABLE OF CONTENTS

50 Things to Know
Book Series
Reviews from Readers
BOOK DESCRIPTION
TABLE OF CONTENTS
DEDICATION
ABOUT THE AUTHOR
INTRODUCTION
1. What is archaeology?
2. Why is archaeology useful/important?
3. What does it take to become an archaeologist today?
4. What was the first known attempt at archaeology in the world?
5. What was the first known attempt at archaeology in the United States of America?
6. Graverobbers or archaeologists?
7. When did archaeology become a formal discipline?
8. Archaeology and The New Deal
9. Archaeology and Cultural Resource Management
10. Tools of the Trade
11. How do archaeologists choose where to dig?
12. Archaeological Excavation

50 Things to Know

13. Soil Discoloration
14. Perishable vs Non-Perishable Material Culture
15. Lithics
16. Ceramics
17. Dating Methods
18. What do archaeologists mean by BP?
19. Archaeological Time Periods
20. Artifact Storage and Care
21. Artifact Restoration
22. Archaeological Funding
23. Repatriation
24. The Great Rift Valley
25. The Neander Valley
26. The Lascaux Cave
27. Australian Rock Art
28. Gobekli Tepe
29. Çatalhöyük
30. The Varna Cemetery
31. Newgrange
32. Otzi the Iceman
33. Skara Brae
34. Stonehenge
35. The First Pyramid
36. The Indus River Valley
37. Bog Bodies
38. The Minoans of Knossos

39. The Acropolis of Athens
40. The Qin Tomb Terracotta Warriors and Horses
41. The Dead Sea Scrolls
42. Pompeii
43. Hadrian's Wall
44. Teotihuacan
45. Sutton Hoo
46. Scandinavian Horse Burials
47. Cahokia Mounds
48. L'Anse aux Meadows
49. Machu Picchu
50. The RMS Titanic
Other Helpful Resources
50 Things to Know

DEDICATION

This book is dedicated to my family.

50 Things to Know

ABOUT THE AUTHOR

As the owner of Musgrave Publishing, Consulting, and Photography, JL Musgrave has currently published twenty books (fiction and non-fiction), created The Hoosier Historian e-magazine, and the What A WanderFull World travel blog, as well as written for other publications such as 812: The Magazine of Southern Indiana, Hektoen International: A Journal of Medical Humanities, and has extensive experience in undergraduate and graduate level academic writing. Musgrave has completed international consultation and review work for authors/writers in Europe, North America, and Australia. Musgrave holds three undergraduate degrees concentrating in social and historical studies, has studied and traveled abroad, and is currently working on an MLitt in the History and Archaeology of the Highlands and Islands. She is passionate about sharing her love of writing, history, travel, cultural anthropology, and archaeology with others through a variety of mediums.

Follow JL Musgrave on Facebook at https://www.facebook.com/JL-Musgrave-152686651819210/

or on Twitter at https://twitter.com/jl_musgrave.

50 Things to Know

INTRODUCTION

"Monuments and archaeological pieces serve as testimonies of man's greatness and establish a dialogue between civilizations showing the extent to which human beings are linked."

Vicente Fox

From the time I was a small child I have loved history, reading, writing, and exploring what this great big world has to offer. Going back to school to earn degrees in history and anthropology only deepened these first loves. Interning at the Glenn A. Black Laboratory of Archaeology and visiting archaeological sites in various parts of the world turned that love into a passion. *50 Things to Know About Archaeology* is a product of this passion to share with novice enthusiasts the real world of archaeology in a simple and concise format.

We all love the adventure and romance of Indiana Jones; however, the real world of archaeology entails so much more. Archaeology allows for an unparalleled glimpse into the past and the material culture that gave birth to human civilization. Through the knowledge gained via archaeological research, we are able to better understand our ancestral roots and the commonalities of humanity that transcend the test of time.

The following is a brief look at the world of archaeology, including frequently asked questions, methods and tools of the trade, and a selection of significant archaeological sites. Utilizing a one paragraph per topic format allows the reader to create an archaeological foundation to be further built upon without overwhelming amounts of technical jargon. By exploring the past, we better understand the present, allowing us to create a brighter future.

1. WHAT IS ARCHAEOLOGY?

Archaeology is the scientific study of humanities' past via the excavation and analysis of bodily and cultural remains. Not to be confused with paleontology, the study of animal and plant fossils, archaeology deals with human interactions within the social and natural world. Archaeology unearths the secrets of the past allowing us a glimpse into the lives of our predecessors.

2. WHY IS ARCHAEOLOGY USEFUL/IMPORTANT?

Archaeology sheds light on the feats and failures of the past. In doing so, it allows modern humanity to learn from our ancestral mistakes and advancements in order to better understand where we come from and where we are going. By studying the lifeways and achievements of the past, we become endowed with the tools to create a better future.

3. WHAT DOES IT TAKE TO BECOME AN ARCHAEOLOGIST TODAY?

To work in the field of archaeology as a field assistant or museum technician, one must possess a minimum of a bachelor's degree. To be a professional archaeologist requires a masters or doctorate. Internships and field schools are highly recommended to acquire hands on experience.

4. WHAT WAS THE FIRST KNOWN ATTEMPT AT ARCHAEOLOGY IN THE WORLD?

There is some debate about the first known attempt at archaeology. Traditionally, the first archaeological endeavor is credited to the Babylonian King Nabonidus (555-539 B.C.E.) via the excavation of an ancient stone foundation. However, others have argued that this honor belongs to Egypt's New Kingdom (1550-1070 B.C.E.) with the unearthing and restoration of the Sphinx.

5. WHAT WAS THE FIRST KNOWN ATTEMPT AT ARCHAEOLOGY IN THE UNITED STATES OF AMERICA?

Thomas Jefferson was America's first scientific archaeologist and as such has been dubbed the Father of American Archaeology. He is given this honor for his excavation work on an Amerindian burial mound in Virginia. By using systematic trenching techniques and observance of stratigraphy, Jefferson was able to conduct the first scientific archaeological dig in America publishing his findings.

6. GRAVEROBBERS OR ARCHAEOLOGISTS?

Throughout the centuries, graves have been plundered from ancient Egyptian pyramids to Native American burial mounds. Early archaeological attempts have been likened to this practice, so what divides the archaeologist from the graverobber? Grave robbing is the practice of plundering a grave for profit, while archaeology is governed by laws that require a permit and excavation of graves for the purpose of scientific research.

7. WHEN DID ARCHAEOLOGY BECOME A FORMAL DISCIPLINE?

Archaeology became a formal discipline in the late 1800s and early 1900s. During this early period, archaeology was generally practiced in Europe, the Americas, Egypt, and "the cradle of civilization" in southwest Asia. Earliest archaeological endeavors were primarily for religious or monetary gain.

8. ARCHAEOLOGY AND THE NEW DEAL

The FERA and WPA were established by President Franklin Delano Roosevelt as the Federal Emergency Relief Association and the Works Progress Administration to supply jobs and income during The Great Depression. Archaeology was a significant part of both the FERA and WPA. Amerindian sites such as Angel Mounds in southern Indiana were excavated as part of the WPA and played a significant role in supplying employment for the region, as well as contributing exponentially to the archaeological record. Upon the conclusion of the WPA, the Civilian Conservation Corps (CCC) continued archaeological excavations until World War II.

9. ARCHAEOLOGY AND CULTURAL RESOURCE MANAGEMENT

Cultural Resource Management (CRM) work entails the preservation and management of material culture from art to a region's heritage. Archaeology plays an important role in CRM work. Cultural Resource Management is organized and executed via government supervision by each country or state. Before a highway or major construction project, CRM workers are brought in to ensure that important artifacts of cultural heritage are not destroyed in the process. If a site is deemed significant, archaeologists are called in to perform an excavation of the site and preserve materials found within.

10. TOOLS OF THE TRADE

Archaeologists use a wide range of tools. Artifacts and other archaeological remains must be extracted without causing further damage or losing research material. The most common tool of the trade is the trowel, much like those used in masonry work, to scrape away layers of dirt. Brushes are used for delicate extraction work. Screens filter through dirt to extract smaller items that would otherwise be lost. Other common archaeological tools include shovels, hand brooms, dustpans, levels, tape measures, string lines, cameras, soil cores, graphing paper, etc.

11. HOW DO ARCHAEOLOGISTS CHOOSE WHERE TO DIG?

Research and preliminary surveillance are key to selecting any dig site. Researching historical records and maps can give insight into what areas may hold interest. Visual surveillance, as well as magnetometry, metal detection, and ground penetrating radar can further narrow the field for excavation.

12. ARCHAEOLOGICAL EXCAVATION

After a site has been selected for excavation, a one meter by one-meter unit is measured out and outlined with string. A shovel is used to remove the top layer of turf. From this point forward trowels are used to evenly remove each layer of dirt. A hand broom and dust pan can be employed to remove excess dirt and keep each unit clean and even. The excavated dirt is transported in buckets to be filtered through screens to remove smaller items. When an artifact is discovered during excavation, brushes can be used to avoid damaging the item. Each item found is recorded via graphing its location within the unit and site's stratigraphy then removed to be analyzed for scientific research.

13. SOIL DISCOLORATION

Not all evidence of material culture can be removed as solid artifacts from a site for examination. Some items leave only discoloration of the soil as evidence of their existence. Examples of this can be found in burial and ceremonial mounds, as well as in places where houses, hearths, post holes, pits, and middens (places where waste was disposed of) once existed. Such discolorations are examined in place, photographed, graphed, and soil or core samples are removed for further examination in a lab.

14. PERISHABLE VS NON-PERISHABLE MATERIAL CULTURE

Most artifacts found by archaeologists are non-perishable items such as lithics and ceramics; however, some perishable items such as food and clothing have been found. Climate and environment play a large role in determining whether a perishable item is preserved. Arid and dry climates or waterlogged environments can preserve otherwise perishable items such as plant, human, and animal-based remains.

15. LITHICS

Lithics are any stone items altered or created by humans. Though most commonly associated with stone tools and weaponry, lithics also covers other stone structures such as megaliths like Stonehenge. Lithics make up the majority of archaeological findings due to their natural longevity and widespread global use.

16. CERAMICS

Ceramics in archaeology consist of pottery, pipes, figurines, and other items made of clay. These items having undergone the firing process, even when broken up into little pieces, tend to stand the test of time. Due to their longevity, like lithics, ceramics are frequently found during archaeological excavations providing information on artistry, religious practices, medicinals, and foodways through scientific analysis of design and residue.

17. DATING METHODS

Archaeologists use various dating methods to ascertain an artifact or site's age. Relative dating uses stratigraphy to determine the order of a series of events; however, it does not establish an exact age. Absolute dating uses methods such as carbon-14 dating and thermoluminescence to ascertain a more precise date range. Most absolute dating methods measure the rate of radioactive decay within an object in one form or another.

18. WHAT DO ARCHAEOLOGISTS MEAN BY BP?

In archaeological terms, BP means "years before the present." When placed after a number such as 2,500 BP, it states that the item is approximately 2,500-years-old. Once radiocarbon dating has established an item's BP, that number can then be calculated to ascertain a date of origin.

19. ARCHAEOLOGICAL TIME PERIODS

Archaeologists and historians divide time into ages or eras based on material culture. Prehistory is divided into three primary ages: the stone age, the bronze age, and the iron age. Each age is divided up according to the primary technological aspects of the time. The stone age is divided further into three eras: the Paleolithic, the Mesolithic, and the Neolithic. Each era of the stone age is divided according to advancements and alterations within stone technology. In North America, the prehistoric period is divided into stages: the Paleo-Indian, the Archaic, and the Formative stage. These are further divided according to region, with varying religious, cultural, and technological adherences as the differentiating factors.

20. ARTIFACT STORAGE AND CARE

Proper artifact storage and care is essential to preserving the longevity of any prehistoric or historic piece. Climate controlled storage helps to control deterioration caused by varying temperature and humidity levels. Each piece is cataloged and numbered for identification purposes, along with any provenance (origin history) information. When handling artifacts, it is crucial to take the proper precautions so as not to cause further damage. Artifact handlers should have clean, dry hands wearing contaminant free gloves when appropriate. Each artifact should be handled with gentle care and its physical condition taken into consideration when assessing levels of contact, display, and preservation methods.

21. ARTIFACT RESTORATION

One form of artifact preservation for display of broken or damaged pieces is restoration. Artifact restoration should be done with care, using minimally invasive tactics. If not done properly using the correct materials and techniques, the artifact can be further damaged or compromised. If done properly, artifacts can be returned to some semblance of their former glory. Restoration training can be acquired through an academic program or apprenticeship.

22. ARCHAEOLOGICAL FUNDING

Archaeological funding comes from a variety of sources. During The Great Depression, government funding was provided through the FERA, WPA, and CCC. Today, funding is provided through scholarships, fellowships, grants, awards, endowments, and donations.

23. REPATRIATION

Repatriation in archaeology is the return of human or cultural remains to their country of origin. In North America, repatriation of Native American burial remains is required by law and governed by the Native American Graves Protection and Repatriation Act (NAGPRA). The process of archaeological repatriation began during the twentieth century when exploited countries and cultural groups requested the return of their cultural resources and ancestral remains.

24. THE GREAT RIFT VALLEY

The Rift Valley in Africa is renowned for its volcanic preservation of early hominid remains. In 1959, paleoanthropologist Mary Leakey discovered the remains of an australopithecine at Olduvai Gorge, Tanzania dated to be 1.8-million-years-old. Many discoveries would follow, including Lucy a 3.18-million-year-old *Australopithecus afarensis* in Hadar, Ethiopia discovered by paleoanthropologist Donald Johanson in 1974. Lucy was named after the song that was playing at the time "Lucy in the Sky with Diamonds" by The Beatles. The region has also been credited for the oldest known stone tools.

25. THE NEANDER VALLEY

In 1856, quarry workers discovered skeletal remains in The Neander Valley, near Dusseldorf, Germany, dating to approximately 300,000 years old. These remains would come to be known as Neanderthal and have been found across Europe ranging in age from 300,000 to 30,000-years-old. Neanderthals are credited with the earliest known ceremonial burial practices.

26. THE LASCAUX CAVE

European cave art is some of the earliest known art in the world. In 1940, four teenage boys discovered a cave with Upper Paleolithic art covering the walls and ceiling near Montignac, Dordogne, France. The Lascaux cave is home to approximately 600 paintings and 1,500 engravings providing the largest number of Paleolithic animal representations yet found. Of particular interest is the Hall of Bulls containing depictions of bulls up to sixteen feet tall. Other Paleolithic cave art have been found throughout France and the Iberian Peninsula.

27. AUSTRALIAN ROCK ART

Indigenous rock art has been found throughout Australia some dating to 50,000-years-old. Australian rock art depicts animals, people, and other shapes painted and engraved into the surface of the rock. Dating of the art is done by analyzing the archaeological layers of occupation in the area.

28. GOBEKLI TEPE

Gobekli Tepe is an 11,000-year-old megalithic temple near Urfa, Turkey. It is believed to be the oldest known temple in the world changing previous notions of the order of civilization development. The pillars that make up the temple are T-shaped, some of which bear the engravings of foxes, birds, lions, scorpions, etc. The various wild animal bones found there indicate that the site was built by hunter/gatherers. Other pillars appear to represent faceless god like figures.

29. ÇATALHÖYÜK

Çatalhöyük, in Turkey, is one of the oldest known Neolithic cities in the world having been occupied from 7500 BCE to 5700 BCE. Evidence of agriculture, animal domestication, art, specialty occupations, and religious observance have been found throughout the site. The city consisted of 1,000 connecting houses, housing approximately 5,000 residents. Upon their deaths, residents of Çatalhöyük were buried in the dirt floors of their familial homes.

30. THE VARNA CEMETERY

The Varna Cemetery is a Late Copper Age archaeological site near Varna, Bulgaria that dates between 4500 BCE to 4000 BCE. It was discovered in 1972 by workmen digging cable trenches. Found within were a wide range of burial goods, indicating a vast trading network, including the oldest known gold objects in the world.

31. NEWGRANGE

Newgrange, in Ireland's Boyne River Valley, is a 5,200-year-old passage tomb built around 3200 BCE by the local farming community over many decades. Intricate circular designs decorate the interior stones of the tomb alongside chambers of human remains. The circular tomb's window aligns with the winter solstice lighting the entire length of the passage. Used for burials and ancestral reverence, the tomb may also have been a ceremonial gathering place for the community.

32. OTZI THE ICEMAN

In the Alps, between Austria and Italy, in 1991, two German hikers discovered a 5,300-year-old corpse buried in the ice. The corpse was removed to a laboratory in Innsbruck, Austria where it was discovered that the cause of death was an arrowhead in his right shoulder. Preservation of the corpse via ice showed that Otzi was killed in the autumn determined by scientific analysis of pollen found with the body. Other injuries to the body showed that the Iceman, dubbed Otzi, had been in a fight with four other people, shown by the four different blood types found on Otzi's weaponry and clothing. Otzi was found wearing a woven grass cape, a goat skin cape, a loin cloth, fur leggings, a hat, and shoes stuffed with grass for warmth. He was also found with a copper ax, knife, bow, arrows, and quiver. An examination of the isotopes in his teeth from the water he had drank over his lifespan revealed that Otzi had lived his entire life in the region. Recent DNA results have revealed that Otzi's familial descendants still live in the region.

33. SKARA BRAE

Skara Brae is a 5,000-year-old Neolithic settlement on the island of Orkney, an island filled with archaeological wonders, just off the northern coast of Scotland. The site was preserved under a sand dune and revealed during a storm in 1850. Sometimes referred to as "the Scottish Pompeii," Skara Brae's level of preservation revealed an interesting glimpse into daily life during the Neolithic period including stone furniture consisting of a dresser/cupboard, bed bases, and water tight storage receptacles.

34. STONEHENGE

Megaliths dot the landscape of The British Isles inspiring myth and wonder, none more so than Stonehenge on the Salisbury Plain in England. The first stage of construction began approximately 5,000-years-ago as an earthwork and timber circle around 3100 BCE and was utilized for religious ceremonies with evidence of human remains. The second stage occurred approximately 1,000 years later in 2150 BCE when bluestones were brought via raft and rollers from the Preseli Mountain in Wales. The third stage around 2000 BCE brought about the addition of the Sarsen stones creating a horseshoe shape. The final stage took place around 1500 BCE when the stones were rearranged into a circle and horseshoe shape.

35. THE FIRST PYRAMID

There are many magnificent archaeological wonders in Egypt that have shaped the way we view the past, from The Pyramids of Giza to King Tut's Tomb in the Valley of the Kings. Every great feat has its beginnings and the same is true of the pyramids. The first pyramid was built in Saqqara by order of King Djoser 4,7000-years-ago using step construction. It has been theorized that the steps may have represented the pathway to heaven. Djoser's chancellor, Imhotep is credited as the architect.

36. THE INDUS RIVER VALLEY

The Indus Valley civilization was established around 3300 BCE in what is now Pakistan and Northern India by the ancient Dravidian peoples. This site is significant as the first civilization of the Indian sub-continent building cities such as Harappa and Mojen-Daro. The Indus Valley civilization was the first to utilize indoor plumbing.

37. BOG BODIES

Imagine going to work on a day like any other day to cut peat from a local peat bog, only to discover a human hand or foot sticking out of the dirt. This scenario and others like it have been happening across Europe for hundreds of years. The oldest known "bog bodies," date back to eight thousand years ago. Bodies of ancient peoples, such as Lindow Man, Grauballe Man, Meenybradden Woman, Borremose Woman, Datgen Man, Tollund Man, Yde Girl, and Windeby Girl (later classified as a boy) have been meeting their untimely demises from the Neolithic to the Middle Ages sacrificed in Celtic religious rites to the sacred waters that would preserve their bodies via the chemical makeup of sphagnum moss. The preserved bodies not only revealed cause of death, but also daily toiletry practices such clothing, hair styles, and homemade hair gel. Other items of value such as gold, silver, copper, iron, and bronze have been found, as well as large containers of butter giving insight into the ancient Celtic world.

38. THE MINOANS OF KNOSSOS

The Minoans occupied the Aegean islands from approximately 2600 BCE to 1400 BCE. The archaeological site of Knossos revealed a seafaring civilization with reverence for the natural world and a thriving trade network. The most dominant site at Knossos is the palace complex and seat of the Minoan civilization. Excavations found colorful images of sea life, birds, flowers, bulls, and people decorating the walls, as well as intricately decorated pottery and inscribed clay tablets. Depictions of sporting events such as bull leaping and boxing were also found among the ruins. It is believed that the Minoans met their end due to the volcanic explosion on the island of Thera.

39. THE ACROPOLIS OF ATHENS

The acropolis in Athens is an ancient citadel that holds some of Greece's most famous archaeological wonders, including the Parthenon, and was first occupied around 1900 BCE. The much-lauded Elgin Marbles, a collection of marble sculptures, once resided within the Parthenon, a temple built from 477 BCE to 432 BCE, and were removed to the British Museum in London, England by Thomas Bruce, 7th Lord of Elgin. The marbles are the subject of continuing dispute for repatriation between England and Greece.

40. THE QIN TOMB TERRACOTTA WARRIORS AND HORSES

The Terracotta Army in China was constructed from 246 BCE to 206 BCE to guard the tomb of China's first Emperor, Qin Shihuang. The project took 720,000 builders to complete. Each warrior is unique with variances in facial features, height, facial hair, and rank.

41. THE DEAD SEA SCROLLS

The Dead Sea Scrolls were created by the Essenes, a Jewish religious sect, from about 200 BCE to 68 CE. They were discovered in 1946-7 by a Bedouin shepherd named Muhammad abh-Dibh. Between 1947 to 1956, 800 different scrolls were found in jars in eleven caves, including a scroll made of copper. Most of the scrolls were written on papyrus and leather containing religious and historic text.

42. POMPEII

Italy is home to a multitude of archaeological wonders from Etruscan tombs to the Colosseum in Rome. Pompeii is a perfect example of volcanic preservation. When the ash fell on Pompeii in 79 CE, it encased the city and its inhabitants preserving their last moments from the positions they fell in to the food they had been eating. Since the site was rediscovered in 1748, Pompeii has provided the archaeological world with a rich source of ancient Roman culture. Many of Pompeii's works of art, including beautiful mosaics and frescoes, are now housed in the Naples National Archaeological Museum.

43. HADRIAN'S WALL

In order to protect their most northern border, the Roman Empire, under the rule of Emperor Hadrian, built a wall between Pictland (Scotland) and Britannia (England) in 122 CE. Spanning from Wallsend on the River Tyne to Bowness on the Solway Firth, Hadrian's Wall was patrolled by Roman soldiers in forts and towns built along the wall. One such fort is Vindolanda where archaeologists found evidence of both Roman and civilian occupation. Items that would normally have been lost to decay such as a woman's wig, a child's sock, sandals, bags, tent fragments, wooden objects, and metal objects, such as coins that would normally have corroded, were found here allowing unique insight into the material culture of the region. Wax tablets in wooden frames with Latin inscriptions were also found, including the earliest known Latin writing by a female.

44. TEOTIHUACAN

The city of Teotihuacan in Mexico was established between 150 BCE and 200 CE. The cities dominant features included two large pyramids and a sacred avenue. Teotihuacan's art, culture, and religion had far reaching effects throughout Mesoamerica with an extensive trading network across the Americas. Archaeological evidence has shown that the practice of human and animal sacrifice was frequently exercised in supplication and appeasement to the gods for blessings in fertility, climate, and agriculture.

45. SUTTON HOO

Sutton Hoo is a series of burial mounds in Suffolk, England. In 1931, archaeologists began excavations and found a trove of rich Anglo-Saxon burial goods. In 1939, they opened the largest mound (Mound 1) and found within its depths evidence of a wealthy ship burial. Later excavations from 1960s to 1980s revealed a prehistoric settlement had once existed under the cemetery. Grave goods from the site consisted of beautiful gold and jeweled accessories, a gilded harness, and armor. Human sacrifices were made at the time and buried in proximity to the wealthier burials. It is believed that the wealthiest burial belonged to an Anglo-Saxon king.

46. SCANDINAVIAN HORSE BURIALS

Horses were an important aspect of Viking age culture in Scandinavia and horse burials can be found throughout Norway, Iceland, Ireland, and Scotland. A horse was not only a beast of burden in agriculture or transportation, but was also a symbol of status. It is as a status symbol that the horse becomes an important figure in religious and funerary practices perhaps inspired by the horse's relationship to the gods Odin and Freyr in Norse mythology and its potency as the figure in many fertility cults.

47. CAHOKIA MOUNDS

Cahokia was built 1000-years-ago and was the largest Amerindian city north of Mexico. Cahokia served as a Mississippian cultural center that covered 4,000 acres with an extensive trade network across the Americas. The prominent feature of Cahokia were the mounds used for burial and ceremonial purposes.

48. L'ANSE AUX MEADOWS

L'Anse aux Meadows was a Viking settlement founded around 1000 CE in Newfoundland, Canada. L'Anse aux Meadows was the first European settlement in the Americas predating Christopher Columbus by nearly 500 years. Excavations in 1960 showed evidence of Viking houses, boats, and an iron forge. Archaeological excavations also showed that the region had been occupied by indigenous persons from approximately 6,000-years-ago to 200 years before Viking settlement. Vikings would continue to use the site whether in a continuous or seasonal capacity for nearly 500 years.

49. MACHU PICCHU

Machu Picchu was built around 1450 CE by the Incas in the Peruvian Andes. Machu Picchu was discovered in 1911 by American explorer Hiram Bingham who believed it was the last Incan refuge from Spanish invaders. Under further scrutiny, Bingham's theory altered to believing Machu Picchu may have been the "birth place" of the Incas. The site included temples, religious shrines, terraces, irrigation canals, and dry stack stone architecture. It is now believed that the site served as a place of royal residence that was abandoned before the conquest of the Spanish.

50. THE RMS TITANIC

On the night of April 14-15, 1912, the RMS Titanic hit an iceberg and sank to the bottom of the Atlantic Ocean. The infamous luxury liner's final resting place was discovered in 1985. Advances in underwater archaeological methods and technology have allowed archaeologists to reveal the secrets of its passengers' last days and opened a new world of underwater exploration.

50 Things to Know

OTHER HELPFUL RESOURCES

- *Archaeology: A publication of the Archaeological Institute of America - https://www.archaeology.org/*
- *Archaeology: Unearthing the Mysteries of the Past* - by Kate Santon
- *The Atlas of World Archaeology* – by Paul G. Bahn

50 Things to Know

READ OTHER 50 THINGS TO KNOW BOOKS

50 Things to Know to Get Things Done Fast: Easy Tips for Success

50 Things to Know About Going Green: Simple Changes to Start Today

50 Things to Know to Live a Happy Life Series

50 Things to Know to Organize Your Life: A Quick Start Guide to Declutter, Organize, and Live Simply

50 Things to Know About Being a Minimalist: Downsize, Organize, and Live Your Life

50 Things to Know About Speed Cleaning: How to Tidy Your Home in Minutes

50 Things to Know About Choosing the Right Path in Life

50 Things to Know to Get Rid of Clutter in Your Life: Evaluate, Purge, and Enjoy Living

50 Things to Know About Journal Writing: Exploring Your Innermost Thoughts & Feelings

50 Things to Know

50 Things to Know

Website: 50thingstoknow.com

Facebook: facebook.com/50thingstoknow

Pinterest: pinterest.com/lbrennec

YouTube: youtube.com/user/50ThingsToKnow

Twitter: twitter.com/50ttk

Mailing List: Join the 50 Things to Know Mailing List to Learn About New Releases

50 Things to Know

www.ingramcontent.com/pod-product-compliance
Lightning Source LLC
Chambersburg PA
CBHW030509220526
45464CB00006B/2720